International Conference on Mechanical Engineering and Applied Mechanics (ICMEAM 2021)

IOP Conference Series: Materials Science and Engineering
Volume 1210

Sanya, China
12 – 14 November 2021

Editor:

Iris Zhang

ISBN: 978-1-7138-4669-7
ISSN: 1757-8981

Printed from e-media with permission by:

Curran Associates, Inc.
57 Morehouse Lane
Red Hook, NY 12571

CURRAN ASSOCIATES INC.
proceedings.com

Some format issues inherent in the e-media version may also appear in this print version.

Institute of Physics
Dirac House, Temple Back
Bristol BS1 6BE UK

Phone: 44 1 17 929 7481
Fax: 44 1 17 920 0979

techtracking@iop.org

Additional copies of this publication are available from:

Curran Associates, Inc.
57 Morehouse Lane
Red Hook, NY 12571 USA
Phone: 845-758-0400
Fax: 845-758-2633
Email: curran@proceedings.com
Web: www.proceedings.com

TABLE OF CONTENTS

IOP Conf. Series: Materials Science and Engineering 1210 (2021) 011001 doi:10.1088/1757-899X/1210/1/011001

International Conference on Mechanical Engineering and Applied Mechanics[ICMEAM 2021], 13 November, 2021

1. About ICMEAM2021

Yaseen Academy organized the **2021 International Conference on Mechanical Engineering and Applied Mechanics (ICMEAM 2021)**, which was planned to be held during 12-14 November, 2021 at Sanya, Hainan Province, China. Due to the restrictions caused by Covid-19, the participants joined the conference online via Tencent Meeting at 13 November, 2021. The Conference looks for significant contributions to all major fields of Mechanical Engineering, Structural Mechanics and Mechanics of Materials in theoretical and practical aspects. The aim of the conference is to provide a platform to the researchers and practitioners from both academia as well as industry to meet and share cutting-edge development in the field.

Organizer: *Yaseen Academy* is located at Huangjiahu Lake University Town, Wuhan, Hubei, China, focusing on international academic conferences organizing and international academic journals corperation in areas of science and engineering, aiming to provide a high-level and open academic platform for global scholars, researchers and practioners. It has gathered an academic committee composed of more than 200 experts and scholars, and has successfully held many international academic conferences. Facing the vast number of experts and scholars, Yaseen Academic is a team with strong professional quality and professional ethics, it has creating a mature, deep and complete academic exchange platform, not only allowing experts and scholars to show and share their latest academic research results and realizing face-to-face academic exchanges, but also allowing the most cutting-edge scientific information, ideas, and viewpoints to be fully communicated and exchanged, which will benefit all mankind.

2. Call for papers

The conference topics include, but are not limited to the following areas:

1)Mechanical Engineering: Mechanical Manufacturing, Vehicle Engineering, Automation, Mechanical And Electronic Engineering, Mechanical Design And Theory, Mechatronics, Manufacturing Automation/Measurement And Control Technology, Industrial Design, Microelectromechanical System Engineering, Mechanical Artificial Intelligence

2)Structural Mechanics: Structural Analysis, Structural Statics, Structural Dynamics, Structural Stability Theory, Structural Fracture And Fatigue Theory, Nanostructures, Structures And Materials, Structures And Models

3)Materials Science: Mechanics of Materials, Advanced Materials, Functional Materials, Metal Materials, Composites, Nanomaterials and Their Applications, Biomaterials, Material Forming And Control Engineering

3. Conference Committee Members

Prof. Ji Wang, Piezoelectric Device Laboratory, School of Mechanical Engineering and Mechanics, Ningbo University, China

IOP Conf. Series: Materials Science and Engineering 1210 (2021) 011001 doi:10.1088/1757-899X/1210/1/011001

Prof. Dr. habil.ing. Genadijs MOSKVINS, State Professor, specialty "Intelligent systems and technologies", Institute of Energetics (IE), Faculty of Engineering (FE), Latvia University of Life Sciences & Technologies (LULST)
Prof. Shouchen Ma, School of Surveying and Land Information Engineering, Henan Polytechnic University
Dr. Ranying Sun, Zhejiang Univeristy of Technology, Hangzhou, China
Dr. Debabrata Singh, Department of CSIT, ITER, SOA(Deemed to be University), BBSR, ORISSA
Dr. Zul-Atfi Ismail, School of Environmental Engineering, Universiti Malaysia Perlis, Malaysia
Dr. Xiaolin Wang, University of Tasmania, Hobart, Australia
Dr. Haijing Wang, School of Information and Electronic Engineering, Zhejiang University of Science and Technology, Hangzhou, China
Dr. Dalin Zhang, University of New South Wales, Sydney, Australia
Dr. S. Sathish, Department of Mechanical Engineering, Hindustan University, Padur, Chennai, India
Dr. Hacene Mellah, Department of Electrical Engineering, Faculty of Technology, Hassiba Benbouali University of Chlef, Algeria
Dr. Ved Pal Singh, Maharshi Dayanand University, Rohtak, India
Dr. Vijayan Gurumurthy Iyer, Entrepreneurship Development Institute of India, India
Dr. Marcelo V. Garcia, Technical University of Ambato, Ambato, Ecuador
Dr. Gang Zhang, Institute of High Performance Computing, A*STAR, Singapore
Dr. Carmen Alexandrache, Faculty of Generale Sciences, Dunarea de Jos University of Galati, Galati, Romania
Dr. Sara Teidj, Faculty of Science and Technology, Moulay Ismail University of Meknes, Morocco
Dr. Victor Koledov, Department of Research and Development, Innowledgement GmbH, Castrop Rauxel, Germany
Dr. Svetlana von Gratowski, Department of Research and Development, Innowledgement GmbH, Castrop Rauxel, Germany

4. Paper Submission

The total count of paper submission was 8. A preliminary review was conducted by the main editor and only the approved papers were sent for peer-reviewed. The peer review for selection of papers was conducted then, each paper was reviewed by 2-4 reviewers(average three) and finally 4 papers were accepted for paper presentation and publication in the conference. The acceptance rate is 50%.

5. Programs

The conference was begun at 13:30, 13 November, 2021 and Assoc. Prof. Jinyang Xu of Shanghai Jiao Tong University hosted the conference. There were 9 participants in total and 2 invited keynote speakers delivered their speech, each speech cost about 35 minutes and 5 minutes for questions&discussion.

IOP Conf. Series: Materials Science and Engineering 1210 (2021) 011001 doi:10.1088/1757-899X/1210/1/011001

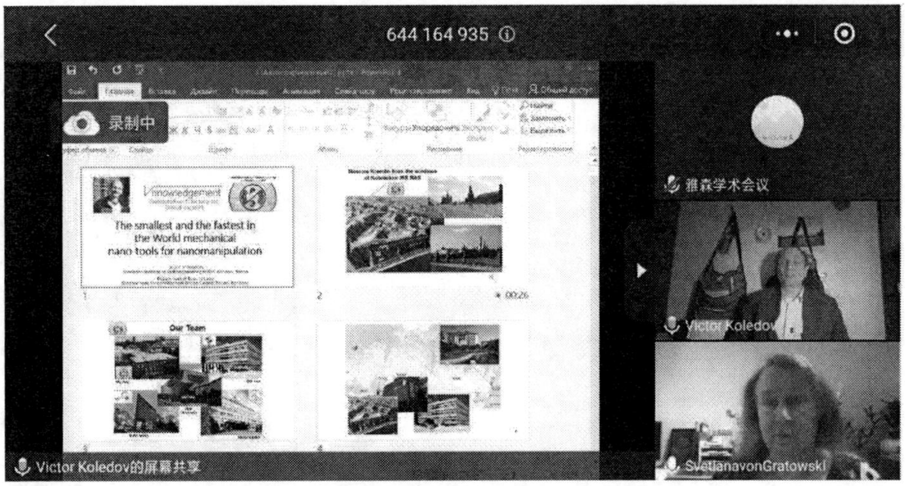

Figure 1: Dr. Victor Koledov is delivering the speech

Speech Title: The smallest and the fastest in the World mechanical nano-tools for nano-mechanic

Dr. Victor Koledov

Department of Research and Development, Innowledgement GmbH, Castrop Rauxel, Germany

Biography: Dr. Koledov Victor was born 20 of April 1955 in city Rostov am Don. From 1972 till 1978 he studied in Study of Moscow Institute of Physics and Technology (MIPT) (Technical University) and finished this University 1978 with Dipl. Eng. in Physics and Electronics. In 1986 he has got PhD (Solid State Physics). In 2008 he has done work for Doctor Habilitation of Sciences with the thesis: "Phase transition and giant deformation in Heusler alloys in external fields.“ (Kotelnikov' IRE RAS). Since 1982 till now he is working Institute of Radio Engineering and Electronics Russian Academy of Sciences (IRE RAS). Now he is Leading scientist in this IRE RAS. His research areas are experimental studies of magnetic properties, magnetooptics, ferromagnetic domains, semiconductor lasers, fiber optics, ferromagnetic shape memory alloys, functional materials, medical applications of functional materials, multiferroics, strong magnetic fields, giant magnetocaloric effect, micromechanics, nanomechanics, nanomanipulation system based on record small nano-tweezers, mechanical bottom up nano-assembling, superconducting magnetic levitation, photonics, nano-optics. He has got awards / Prizes / Honours at 2009 Start2grow competition of Businessplan Germany for Nano-tools, 2010 Start2grow competition of Businessplan Germany for MST-technology of Nano-tools. And after this since 2010 he is founder and director of German innovative Innowledgement GmbH. He is leader of more the 25 national and international projects. Victor Koledov is author and coauthor of more then 180 scientific paper in referred journals and more then 160 presentations at national and international presentation, among then more than 25 Invited and Keynote speaker. He was invited Professor in – Spain, Romania, Portugal, Poland, China, India, Brazil, South Africa. He participates in organization of conferences 1. Conference on Manipulation, Automation and Robotics at Small Scales (MARSS). 2. Conference 3 M nano International Conference on Manipulation, Manufacturing and Measurement on the Nanoscale 2014, 2018 Nano 3 M Award for invited report, 2018 3M Nano Award as Keynote speaker.

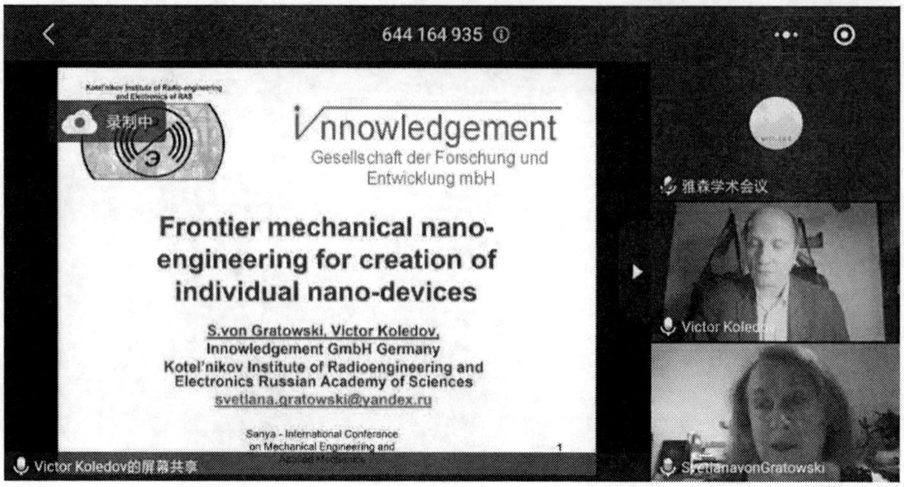

Figure 2: Dr. Svetlana von Gratowski is delivering the speech

Speech Title: Frontier mechanical nano-engineering for creation of individual nano-device

Dr. Svetlana von Gratowski
Department of Research and Development, Innowledgement GmbH, Castrop Rauxel, Germany

Biography: Dr. Sci. Svetlana von Gratowski was born at 22 of May 1955 in Moscow region. From 1972 till 1974 he studied in Study of Moscow Institute of Physics and Technology (MIPT) (Technical University) and from 1974 till 1979 in Moscow State University after Lomonosov. She had finished this University 1979 with Dipl. Physics. From 1979 till 1984 she was in post at the post graduate courses in Institute of Radio Engineering and Electronics Academy of Science USSR. In 1984 she had got degree Dr. of Science (Solid State Physics). Since 1982 till now she is working Institute of Radio Engineering and Electronics Russian Academy of Sciences (IRE RAS). Now she is scientist researcher in this IRE RAS. Her research areas are theoretical physics in magnetism, magnetic domains, electromagnetic waves properties of condensed matter, spin waves, superconductivity, phase transitions, shape memory alloys, functional materials, multiferroics, ferroelectrics, magnetic materials, carbon nanotubes, nano-optics, nano- photonics, nano-plasmonics, micro- and nano-mechanics, nano-tools, nano-manipulation, nano-assembling, bottom up mechanical nano-manufacturing using record small nano-tweezers, non-destructive non-nondestructive and non-invasive real time monitoring of moisture content in plant, - in work of art, -glucose content in blood. She has got awards / Prizes / Honors at 2009 Start2grow competition of Business plan Germany for Nano-tools, 2010 Start2grow competition of Business plan Germany for MST-technology of Nano-tools. And after this since 2010 till now she is founder and deputy director of German innovative Innowledgement GmbH. She is principal investigators of more the 20 national and international projects and Leader one project. Svetlana von Gratowski is author and coauthor of more than 90 scientific paper in referred journals and more, then 160 presentations at national and international presentation, among then more than 11 Invited and Key Note speaker. She was invited researcher in – Italy, Spain, Romania, Portugal, China, India, Vietnam, Brazil, South Africa. She is member of reviewer board in more than 10 international scientific journals, and member of editor's board in 2 international scientific journals. 2014, 2018 she had got 3 M Nano Award for invited report.

IOP Conf. Series: Materials Science and Engineering 1210 (2021) 011002 doi:10.1088/1757-899X/1210/1/011002

Peer review declaration

All papers published in this volume of IOP Conference Series: Materials Science and Engineering have been peer reviewed through processes administered by the Editors. Reviews were conducted by expert referees to the professional and scientific standards expected of a proceedings journal published by IOP Publishing.

- **Type of peer review:** Double-blind review
First of all, the editor Iris will take a simple check to know whether the paper topics are included in the call-for-paper areas of this conference, if not, the submission will be rejected or seggested to other conferences. The included submissions will then be numbered and sent to the conference committee members(reviewers) for peer-review, the author names, emails and affiliations were removed constantly in this porocess. Each reviewer should fill in the reviewing form and 9 criterias in total are to be scored (5-point scale, with 1=lowest and 5=highest), including "Originality: Is the work relevant and novel", "Motivation: Does the problem considered have a sound motivation", "Repetition: Have significant parts of the manuscript already been published", "Length: Is the content of the work of sufficient scientific interest to justify its length", "Conclusion: Does the paper contain a carefully conclusion, summarizing what has been learned and why it is useful", "Reference: Is the reference adequate and necessary", "Structure: Is the manuscript structure(including the title, abstract and the content) rational", "Language: Is the language accurate and readable", "Diagrams, figures, tables and captions: Are they essential and clear". Then the reviewer will give an overall evaluation or Detailed reviewing result(if any). In the end of the form, the final decision of Accept/Revision/Reject should be selected. No matter which selection the paper was marked, we send the whole reviewing form to the authors. When it is marked by Revision, we will have a double check of the revised papers before acceptation.
- **Conference submission management system:** Yaseen Academy

- **Number of submissions received:** 8

- **Number of submissions sent for review:** 7

- **Number of submissions accepted:** 4

- **Acceptance Rate (Number of Submissions Accepted / Number of Submissions Received X 100):** 50%

- **Average number of reviews per paper:** 3

- **Total number of reviewers involved:** 12

- **Any additional info on review process:** A preliminary review was conducted by the main editor and only the approved papers were sent for peer-reviewed.

- **Contact person for queries:** Iris Zhang (Yaseen Academy, Hongshan District, Wuhan, Hubei, China, iriszhang@yaseen.org.cn)

Research on Low Frequency Sound Insulation Properties of Membrane-Type Acoustic Metamaterials

Xiaokai Yin[1], Yongchao Xu[2] and Hongyu Cui[1,*]

[1]School of Naval Architecture and Ocean Engineering, Dalian University of Technology, Dalian, Liaoning 116024, China
[2]Marine Design and Research Institute of China, Shanghai 200011, China

[*]corresponding author's e-mail: cuihongyu@dlut.edu.cn

Abstract. To solve the problem of low-frequency noise control in ship cabins, a new membrane-type acoustic metamaterial (MAM) with bulges on the surface of thin films is designed based on the characteristics of lightweight and low-frequency sound insulation of membrane-type acoustic metamaterials. The sound structure coupling module of COMSOL multiphysical field coupling software is used to analyse the sound insulation performance of MAMs. The sound insulation properties of the additional mass film and self-similar fractal convex structure are further discussed. The metamaterial structure studied in this paper has a better sound insulation effect than ordinary film, which provides strong technical support for ship cabin noise control.

1. Introduction

The radiated noise of a ship not only interferes with the work of the ship's detection system and reduces the ship's acoustic stealth performance, but also causes serious harm to the health of the people on board. Due to its long wavelength, slow attenuation, and longer propagation distance in the water, it is very important to achieve effective control of low-frequency ship noise.

Traditional sound insulation materials follow the law of mass action and often achieve better sound insulation effects by increasing material thickness and areal density. However, this method bears the cost of increasing the weight of the material, which is difficult to achieve for the noise protection design of a ship that requires a lightweight structure and a compact space. Because the structure of acoustic metamaterials breaks through the limitation of the law of mass action, special thin-film acoustic metamaterials have the advantages of light weight, small size and flexible layout, and exhibit good sound insulation performance in the low frequency range, which has become a hot research topic [1-2]. In 2008, at Hong Kong University of Science and Technology, Yang et al. proposed a lightweight MAM. Through theoretical research and experimental verification, it was found that this metamaterial has better low-frequency sound insulation band gap characteristics [3]. Subsequently, researchers studied the structure of thin-film metamaterials, designed various acoustic metamaterials with different structures, and discussed their sound insulation properties [4-6]. Because the MAM has excellent performance in the field of low-frequency sound insulation and vibration reduction, it provides a new way to solve the problem of low-frequency vibration and noise reduction of ship structures.

In this paper, based on ordinary MAMs, a new type of thin-film metamaterial structure is designed by adding convex structures on the surface of the membrane, and the change law of the sound transmission loss (STL) of the metamaterial under different structural parameters is discussed.

2. Samples and methods

2.1. Cell structure design

As shown in figure 1, the film is a square with a side length of l and a thickness of t_l, to add a well-shaped bulge on the membrane with a side length of a and a height of h.

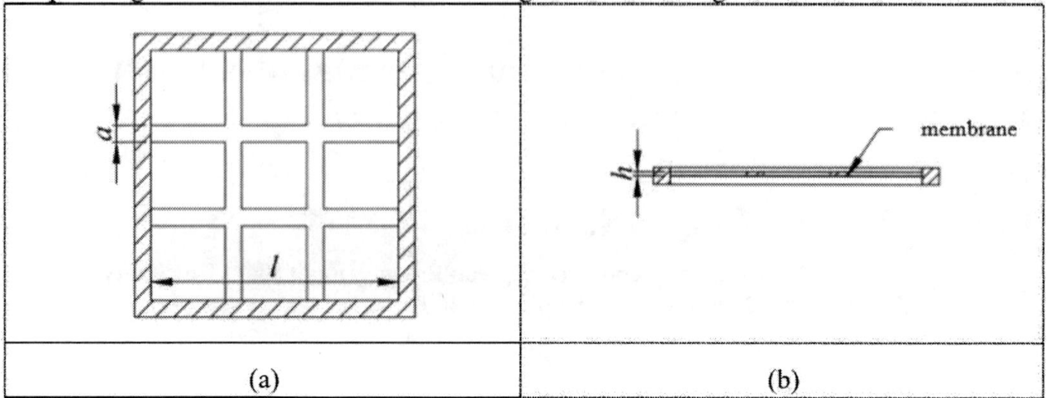

Figure 1. Mode of cell structure. (a) Top view of the cell; (b) Side view of the cell.

2.2. Calculation method

According to the structural size and parameter characteristics of the MAM, the modal superposition method is used to calculate the sound insulation performance of the structural unit [7]. The equation of motion of the film structure with protrusions on the surface is

$$\rho_s \frac{\partial^2 \omega}{\partial^2 t} + \rho_{mass} h\left(x, y, x_0, y_0, l_x, l_y\right) \frac{\partial^2 \omega}{\partial t^2} - T\nabla^2 \omega = p_i + p_r + p_t \tag{1}$$

Where ρ_s, ρ_{mass}, T, p_i, p_r, p_t represent the surface density of the film, the surface density of the convex mass, the film tension, the incident sound pressure, the reflected sound pressure, and the transmission sound pressure.

$$h\left(x, y, x_0, y_0, l_x, l_y\right) = \left[\mathrm{H}(x - x_0) - \mathrm{H}(x - x_0 - r)\right]\left[\mathrm{H}(y - y_0) - \mathrm{H}(y - y_0 - r)\right] \tag{2}$$

$$\nabla^2 = \frac{\partial^2}{\partial x^2} + \frac{\partial^2}{\partial y^2} \tag{3}$$

where H is the Heaviside function (step function), and $\omega(x, y, z)$ is the lateral displacement of a point (x, y) on the film at time t [8].

According to the modal superposition theory,

$$\omega(x, y, z) = \sum_{n=1}^{N} W_n(x, y) q_n(t), \quad q_n(t) = \tilde{q}_n e^{j\omega t} \tag{4}$$

where W_n and q_n are the mode shape and coordinates.

Integrating the entire membrane can obtain the vibration equation of the membrane under the action of acoustic waves, as shown in Equation (5).

$$-\omega^2 \{[M] + [Q]\}\{\tilde{q}\} + j\omega[C][\tilde{q}] + [K][\tilde{q}] = 2\tilde{A}\{H\} \tag{5}$$

Solving Equation (5) can obtain

$$\{\tilde{q}\} = \frac{2\tilde{A}}{-\omega^2 \{[M] + [Q]\} + j\omega[C] + [K]}\{H\} \tag{6}$$

Substituting $L_x = L_y = l$ into the transmission coefficient t_p can be expressed as

$$t_p = \left| \frac{j2\rho_0 c_0 \omega}{l^2} \{H\}^T \frac{1}{-\omega^2 \{[M]+[Q]\} + j\omega[C]+[K]} \{H\} \right| \tag{7}$$

Then, the expression of the STL of the film-type acoustic metamaterial is

$$STL = 20\log_{10}\left(1/t_p\right) \tag{8}$$

3. Result analysis

3.1. STL of MAMs

This study uses COMSOL Multiphysics 5.4 finite element software to numerically simulate the sound insulation of the metamaterial cell structure. The silicone rubber film (l=30mm, t_l=0.2mm) was applied at a 200N/m pre-pressure. The silicone rubber well-shaped bulge (a=4mm, h=0.5mm) was firmly attached to the membrane with a quality of 0.41g. The frame was made of aluminium, and the corresponding material parameters of each material are shown in table 1. The finite element model is composed of MAM elements and air domains on both sides. Among them, the solid frame structure is surrounded by fixed constraints, and a displacement constraint is set between the membrane and the solid frame. One side of the air domain is set as the background pressure field, and the sound pressure amplitude is set to 1 Pa, which is perpendicular to the surface of the film. The outermost part of the air domain is set as a perfect matching layer to eliminate the influence of sound wave reflection on the result. The remaining peripheral boundary conditions are uniformly set as hard sound field boundary conditions. The calculated sound insulation curve of the metamaterial structure in the range of 50–2000Hz is shown in figure 2.

Table 1. The material parameter characteristics.

Parameters	Material	Density($kg \cdot m^{-3}$)	Elastic Modulus(Pa)	Poisson's ratio
Film	Silicone Rubber	980	2×10^5	0.49
Frame	aluminum	2730	6.9×10^{10}	0.33

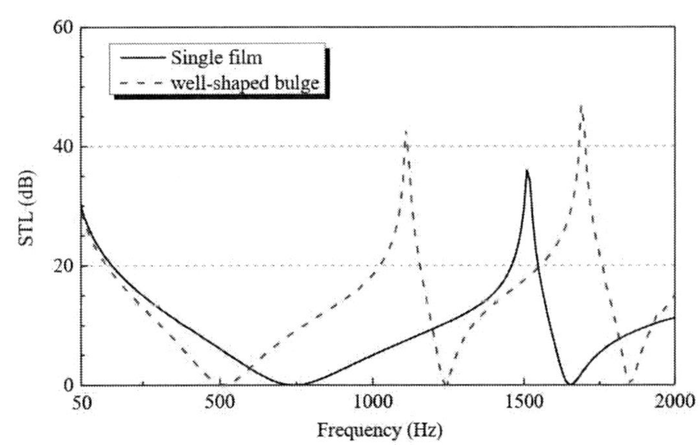

Figure 2. STL curves of metamaterial

Figure 2 shows that adding a convex structure on the surface of the ordinary film can improve the sound insulation effect of the structure in the low frequency range. The ordinary membrane structure has an STL peak at approximately 1500Hz, and the maximum STL is 36dB. The metamaterial structure designed in this paper has two STL peaks near 1200Hz and 1800Hz, and the maximum STL reaches 46dB, which greatly improves the low-frequency STL. Due to the light weight of the membrane structure itself, the metamaterial designed in this paper can achieve a good low-frequency sound insulation effect without sacrificing quality.

3.2. The effects of the membrane properties and bulge configurations

Among the MAMs, the elastic membrane is an important part that affects the sound insulation effect of the material. Therefore, to study the change law of the sound insulation of the MAM, the membrane pre-pressure and the thickness of the membrane were changed to perform numerical calculations. The calculated sound insulation curve of the metamaterial structure is shown in figure 3.

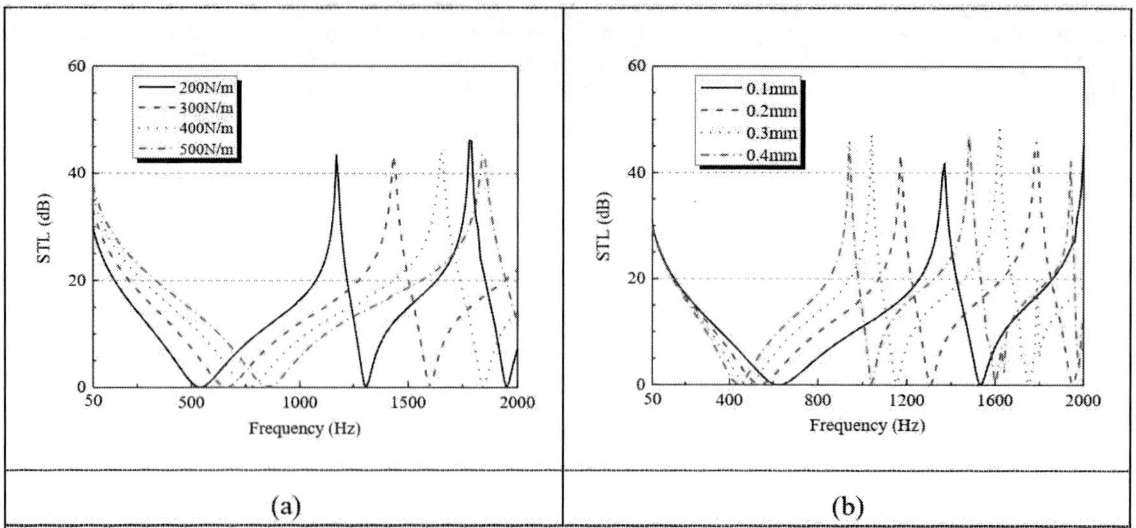

Figure 3. Relationship of STL with film parameters. (a) STL curves of various membranes' prestress; (b) STL curves of various membranes' thickness.

Figure 3(a) calculates the STL curves of the film when the film prestress gradually increases from 200N/m to 500N/m. As shown in figure 3(a), with the gradual increase in the prestress of the film, the centre frequency corresponding to the STL peak of the material moves to high frequency, and the sound insulation capacity and the STL bandwidth gradually increase. Figure 3(b) explores the influence of the film thickness on the sound insulation performance of the film when the film thickness is gradually increased from 0.1mm to 0.4mm. With the gradual increase in the film thickness, the entire STL volume curve moves in the low frequency direction, and the STL volume gradually increases. Therefore, by increasing the film thickness, a better low-frequency sound insulation effect can be obtained, and in practical applications, the film weight is lighter, and increasing the film thickness does not significantly increase the quality of the metamaterial.

The convexity of the film surface designed in this paper is equivalent to adding mass to the surface of the film, so the quality and distribution of the convex structure also affects the sound insulation performance of the metamaterial. To further study the change law of the sound insulation of the MAM, the quality and distribution of the protrusions on the surface of the membrane were changed to perform numerical calculations. The calculated STL curve of the metamaterial structure is shown in figure 4.

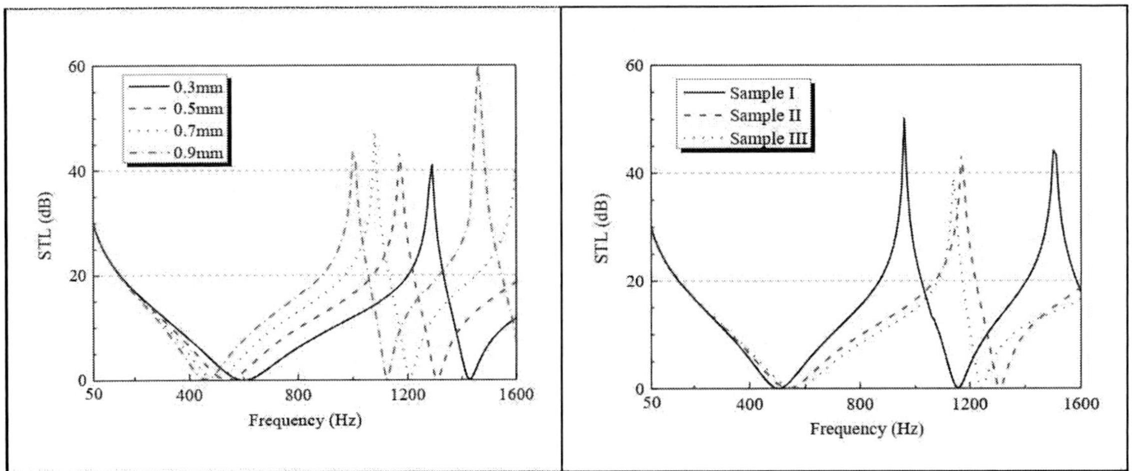

(a) (b)

Figure 4. Relationship of the STL with the parameters of the bulge. (a) STL curves of various bulges' thickness; (b) STL curves of various bulge distributions.

Figure 4(a) explores the influence of the quality of the bulge on the sound insulation performance of the acoustic metamaterial. The height of the bumps is gradually increased from 0.3mm to 0.9mm. Comparing the curves in the figure shows that as the quality of the bulge increases, the STL curve of the material shifts in the low-frequency direction. Figure 4(b) explores the influence of different distributions on the sound insulation performance of the metamaterial structure without changing the quality of the bulges. The basic structure is shown in figure 5. Comparing the STL curves shows that under the same quality condition, the sound insulation effect of Sample I is better. There are two STL peaks near 960Hz and 1500Hz, the maximum STL reaches 50.2dB, and the STL bandwidth is significantly increased.

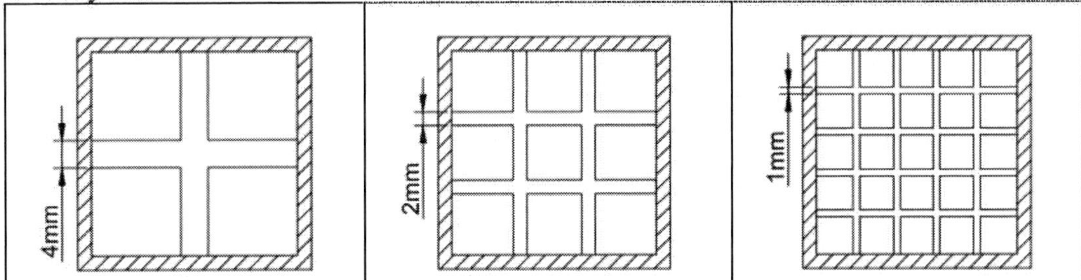

(a) Sample I (b) Sample II (c) Sample III

Figure 5. Basic structure of the film surface bulge under different distributions

3.3. STL of additional mass film

This section studies the sound insulation characteristics of the additional mass film. To add the mass distribution on the membrane, a 2mm-thick mass block with a diameter of 5mm was attached to the membrane. It is made in a neodymium iron boron magnet with a density of 7400kg/m^{-3}, a modulus of elasticity of 1.61011Pa and a Poisson's ratio of 0.28. The different placement positions of the mass blocks are shown in figure 6(a), and the calculated sound insulation curve of the metamaterial structure is shown in figure 6(b).

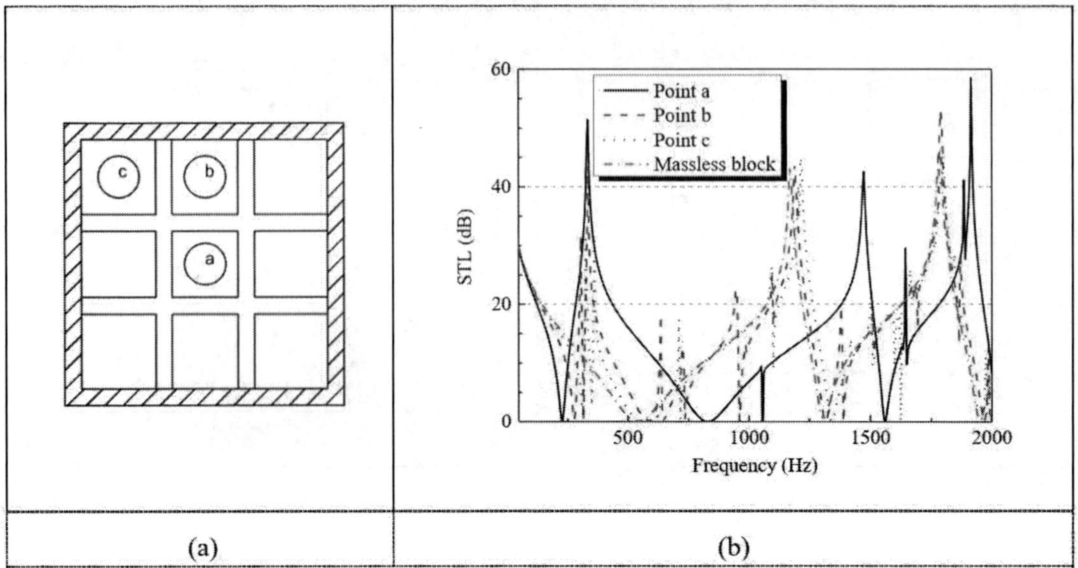

Figure 6. Basic structure and STL curves of the additional mass film

The tic-tac-toe structure divides the film into nine regions. The three points a, b, and c are the intersection points of the diagonals of the respective rectangular regions, and point a is the centre point of the film. As shown in figure 6(b), the first STL peak without a mass block appears near 1170Hz, and the maximum STL is 43.5dB. When the mass is added, a new STL peak appears at 330Hz, and the maximum STL reaches 51.5dB, which greatly improves the low-frequency sound insulation effect of the metamaterial. In addition, when the mass is located at centre point a of the film, the STL bandwidth and STL peak of the metamaterial at approximately 330Hz are optimal, and the frequencies corresponding to the second and third STL peaks move to high frequencies. When the mass block is located at two points b and c, except for the addition of a new STL peak at 330Hz, the STL curve above 500Hz basically coincides with that without a mass block. Therefore, it can be concluded that in the low frequency range below 500Hz, the additional mass film has better sound insulation performance than the massless film, and the metamaterial has the best low frequency sound insulation effect when the mass is located at the centre point.

3.4. Design of convex structure based on Hilbert fractal curve

The fractal structure has multiscale self-similarity, which can effectively broaden the frequency band of metamaterials and improve space utilization. The Hilbert fractal curve is the most widely used and ideal fractal structure in acoustic metamaterials. Figure 7 shows the basic structure of the first- to third -order film surface protrusions based on the Hilbert fractal design. To ensure the constant quality of the bumps, the widths of the first- to third-order bump structures are designed to be 4 mm, 2 mm, and 1 mm. The calculated sound insulation curves are shown in figure 8.

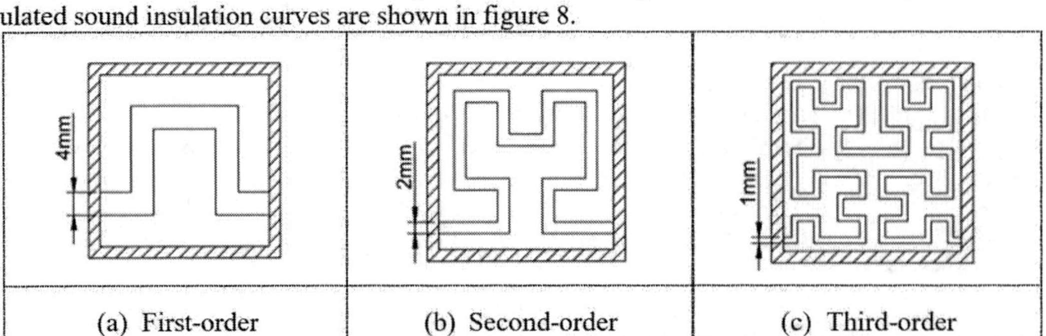

Figure 7. Basic structure of thin film based on Hilbert fractal

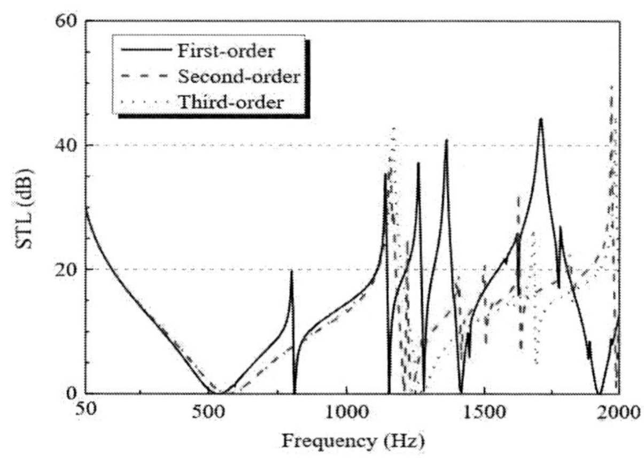

Figure 8. STL curves of thin film based on Hilbert fractal

Comparing the STL curves of the fractal structure and the ordinary well-shaped structure shows that the fractal structure has more STL peaks in the same frequency range. Due to the self-similarity of the Hilbert fractal curve, the structure has more resonance modes. Comparing the first-order, second-order, and third-order STL curves of the fractal structure under three different orders shows that the second-order fractal structure has the most STL peaks, but the peak is smaller, and it cannot produce a better sound insulation effect. The first-order fractal has four STL peaks in the frequency range of 1,000–2,000Hz, and the STL peaks basically reach 40dB, which has a good broadband sound insulation effect. This self-similarity fractal structure can provide a new idea for broadening the frequency band of acoustic metamaterials.

4. Conclusion

Based on the lightweight and low-frequency sound insulation characteristics of MAMs, a new type of MAM with protrusions is designed that could be used for noise protection in ship cabins. The acoustic-structure coupling module of COMSOL software is used to analyse the influence of the metamaterial structure and its parameter changes in film prestress, film thickness, protrusion quality and protrusion structure distribution on its sound insulation performance. The STL performance of the additional mass film and the self-similar fractal convex structure are further discussed. The results of this paper provide powerful technical support for ship cabin noise control.

References

[1] Bolton J S, Shiau N M and Kang Y J 1996 Sound transmission through multi-panel structures lined with elastic porous materials. *Journal of Sound and Vibration*, **191(3)** 317-347.
[2] Wu J, Ma F, Zhang S, et al. 2016 Application of acoustic metamaterials in low frequency vibration and noise reduction. *Journal of Mechanical Engineering* **52(13)** 68-78.
[3] Yang Z, Mei, Yang M, et al. 2008 Membrane-Type Acoustic Metamaterial with Negative Dynamic Mass *Physical Review Letters* **101(20)** 204301.
[4] Cui H, Du Y and Wang H 2021 Analysis of sound insulation characteristics of acoustic metamaterial structures with different shaped masses Ship science and technology **43 (05)** 33-36.
[5] Guo L, Zhang J and Zeng H. 2020 The numerical study of sound Insulation performance of membrane acoustic metamaterial cell with variable thickness. Ship science and technology **42(01)** 38-42.
[6] Lin G, Chen S, Yuan X and Tan H 2016 Transmission loss of thin membrane-type acoustic metamaterials with multi-masses *Int. Conf. on Innovative Material Science and Technology (IMST2016)* (Shenzhen, China)

[7] Zhang J, Yao H, Du J, et al. 2016 Analysis of the Sound Insulation Properties of Membrane-type Acoustic Metamaterial Plate *Journal of Synthetic Crystals* **45(10)** 2549-2555
[8] Zhang Y, Wen J, Xiao Y, et al. 2012 Theoretical investigation of the sound attenuation of membrane-type acoustic metamaterials *Physics Letters A*, **376(17)** 1489-1494.

IOP Conf. Series: Materials Science and Engineering 1210 (2021) 012002 doi:10.1088/1757-899X/1210/1/012002

Structural Optimization of Road Fully Enclosed Sound Barrier Based on Three-Dimensional Finite Element Analysis

Yang jingqiang[1,*], Din dewei[1] and Yang zhimian[2]

[1]CCCC First Harbor Engineering Co. Ltd, Tianjin, China
[2]School of transportation and logistics engineering, Wuhan University of technology, Wuhan, Hubei, China

Email: 2321547575@qq.com

Abstract. With the help of the finite element analysis software ANSYS, this paper establishes a finite element model of a fully enclosed sound barrier of elevated road, calculating and analyzing the stress of supporting structure under different working conditions, checking the strength, stiffness and stability of the structure under the most unfavorable working condition. Finally we optimize the structure according to its characteristics.
Keywords. Fully enclosed sound barrier; ANSYS software; Bearing capacity analysis; Structural optimization

1. Introduction

Since entering the 21st century, transportation has been developing rapidly in cities, but it follows the traffic noise problems[1]. The sound barriers have advantages of obvious noise reduction effect, flexible building processing, small land occupation and easy maintenance[2]. Therefore, setting sound barriers on both sides of the road to control traffic noise is a method that has been widely applied and praised at home and abroad.

For the roads with high-rise sound sensitive buildings along the line, the fully-enclosed sound barriers have obvious advantages over the semi-enclosed sound barriers[3]. The former can increase the sound insertion loss and finally effectively reduce the sound pressure level at the sound receiving point of high-rise floors.

Most of the sound barriers have the characteristics of light weight and large windward surface, they are more significantly affected by the wind than ordinary buildings. Compared with semi-enclosed sound barriers, fully-enclosed sound barriers have a closed top, which intensifies the adverse impact of snow load. Therefore, under the limit state of bearing capacity, the barrier structure shall not show excessive stress or deformation that may pose a threat to vehicles passing inside and make people feel uneasy. From what has been discussed above, the barriers should have the ability to resist various natural disasters such as wind, snow and earthquake, so it is very necessary to calculate and analyze the structure[4].

2. Model building

Based on the single lane ramp of urban elevated road and combined with existing engineering examples, this paper designs the following sound barriers[5]: the fully enclosed sound barrier adopts the form of portal frame, with a total length of 120m, a clear height of 5m, a maximum height of 5.6m and a width of 8m.The steel structure is used as the supporting structure, a portal steel frame is set every meter, and the top purlin is erected on the top to connect and bear the weight of the top plate. As the sound barrier itself is a wind sensitive building, H-section steel with strong bending bearing capacity is selected as the steel frame column. The steel brand is Q235, and the section size is 250mm × 250mm × 9mm × 14mm。Welding structure is adopted between columns and between columns and panels; The bottom of two rows of vertical columns is fixed with concrete.

Since the length of the sound barrier is 120m, if the full-length barrier structure is selected as the research object, the amount of analysis will be too large, which is not conducive to calculation. According to the structural characteristics of the sound barrier, it can be regarded as an infinite body. We select a 12m long sound barrier as the research object, imposing boundary conditions, and considering the influence of other parts on the selected section. The finite element model is shown in Figure 1.

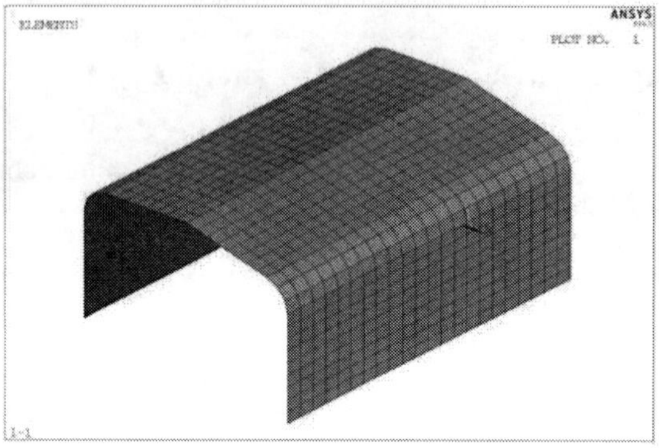

Figure 1. Finite element model of fully enclosed sound barrier

In ANSYS software, beam188 element is used to simulate steel frame and purlins, and shell63 element is used to simulate panels[6]. The element material parameters are shown in Table 1.

Table 1. Element material parameters

Component name	columns and purlins	Sound absorption and insulation panel
Material	Q235 steel	Aluminum alloy composite plate
Element type	Beam188	Shell63
Elastic modulus	2.06×10^{11}Pa	7.2×10^{10}Pa
Poisson's ratio	0.3	0.33
Density	$7.85 \times 10^3 kg/m^3$	$2.8 \times 10^3 kg/m^3$

3. Calculation and analysis

The structural stress calculation and analysis of the fully enclosed sound barrier includes the effects of self weight, wind load, snow load and seismic load. Different loads are applied through ANSYS software to calculate the axial force, shear force, bending moment and stress value distribution of the structure under various working conditions and analyze the most unfavorable position and the safety performance of the barrier structure. The load cases are shown in Table 2.

Table 2. Load cases

working condition	Load combination
Condition I	Dead weight
Condition II	Dead weight + snow load
Condition III	Dead weight + wind load
Condition IV	Dead weight + snow load + wind load
Condition V	Dead weight + seismic load

The structural displacement and deformation under various working conditions and the positions needing attention can be calculated by ANSYS software as shown in Table 3.

Table 3. Summary of working conditions

working condition	Maximum displacement（mm）	Maximum displacement position	Other positions needing attention
Condition I	4.165	Middle of roof	Curved panel
Condition II	5.889	Middle of roof	Curved panel, connection between curved panel and side plate
Condition III	4.912	Windward top plate close to the middle	Connection between curved panel and side plate
Condition IV	5.806	Windward top plate close to the middle	Connection between top plate and curved plate, bottom of steel frame column
Condition V	4.313	Middle of roof	Connections between top and curved plates

It can be seen from table 3 that under condition 2, i.e. snow load, the maximum displacement is in the middle of the roof which is 5.889mm.According to the stress of the barrier structure, condition 4, i.e. wind and snow combined load, is the most unfavorable condition, and the displacement in the middle of the top plate is 5.806mm.The reason why the displacement in the middle of the roof in condition 4 is less than that in condition 2 is that for the double slope roof, the wind load will produce a certain suction on the roof on the windward side, which reduces the vertical displacement of the roof to a certain extent.

Under the combined wind and snow load, the displacement nephogram of condition 4 is shown in Figure 2. The maximum displacement is 5.806mm, which is near the middle of the roof on the windward side. The displacement of the windward side plate is very small, almost 0;The maximum displacement of the side plate on the leeward side is 2.58mm, which is at the connection between the curved plate and the side plate.

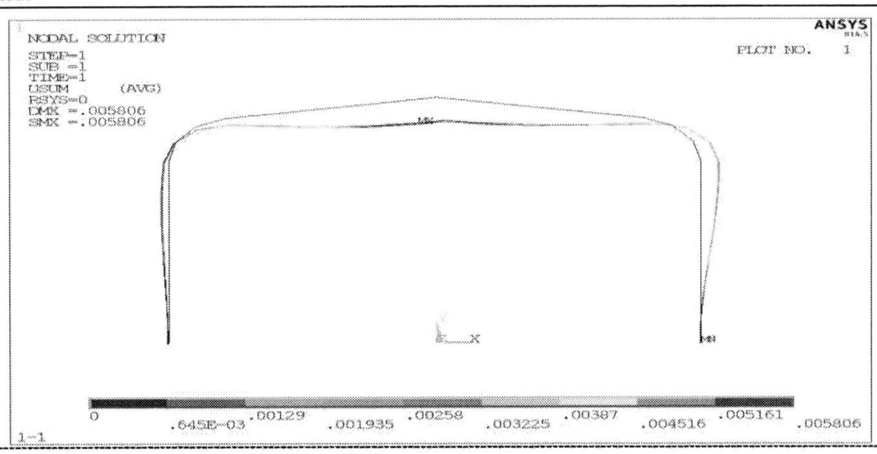

Figure 2. Cloud diagram of displacement under working condition IV

4. Structural checking calculation

The checking calculation of the bearing capacity limit state of the supporting steel frame under the most unfavorable working condition mainly includes three aspects: strength, stiffness and stability.

According to the results obtained by ANSYS software, the maximum compressive stress appears at the bottom of the steel frame column, and the maximum compressive stress is 1298400pa.The maximum tensile stress appears on the cross beam at the connection position between side plate and curved plate, and the maximum tensile stress is 283120pa.The maximum compressive stress and maximum tensile stress are far less than the design value of compressive and tensile strength of Q235 steel 215000000pa.The strength checking calculation meets the requirements. The stress table of steel frame column is shown in Table 4. The negative sign indicates compressive stress, otherwise it is tensile stress.

Table 4. Stress of steel frame column (Pa)

Element number	Section stress
441	-1 298 400
442	-1 100 000
443	-990 360
444	-918 830
445	-863 730
446	-816 940
447	-771 160
448	-720 130
449	-670 340

The stiffness of compression bending members is usually determined by slenderness ratio λ to measure. The smaller the slenderness ratio, the greater the stiffness of the member, and vice versa. Calculated slenderness ratio of column $\lambda = 14.58$, less than 150 allowable slenderness ratio of compression member column in "Code for design of building steel structures" (GB 50017-2003);The slenderness ratio of top purlin and beam is calculated $\lambda = 34.99$, less than the allowable slenderness ratio of tie bar of tension member 400.The stiffness checking calculation meets the requirements.

The criterion to avoid local instability of members is to make the local buckling critical stress greater than the yield strength of steel or the overall stability critical stress of members. In fact, the requirement of ensuring local stability is realized by limiting the width thickness ratio of plates. After calculation, the width thickness ratio of both flange and web are less than the specification limit. The stability checking calculation meets the requirements.

5. Structural optimization

5.1. Column section optimization

The column of barrier steel frame is mainly under pressure, and the maximum stress value is less than the compressive design strength of Q235 steel. Therefore, the section can be optimized. Considering the overall stability of the column, in order to avoid flexural-torsional buckling, the solid-web compression members generally adopt biaxial symmetric section.

The specific steps for designing sections are as follows:

1) Determine the size of the cross-sectional area: assume the slenderness ratio of the member λ is 100. According to the classification of axial compression members in the "Technical code for steel structures of high rise civil buildings" (JGJ 99-1998), it is found that when the plate thickness t of H-section steel is less than 80mm, the x-axis is class B section and the y-axis is class C section, which can be obtained by referring to the attached table $\varphi_x = 0.581$, $\varphi_y = 0.483$. $\varphi_{min} = \varphi_y = 0.483$. It can be calculated that $\mathrm{f} = 215\mathrm{MPa} = 215\mathrm{N/mm^2}$。 Then the required cross-sectional area is calculated as follows:

$$A = \frac{N}{\Phi f} \tag{1}$$

N — the force added to the section;

Φ — Overall stability coefficient;

f — Steel compression strength design value.

IOP Conf. Series: Materials Science and Engineering 1210 (2021) 012002 doi:10.1088/1757-899X/1210/1/012002

The results of Chapter III are preliminarily brought into the available data$A = 17.76\text{cm}^2$。

2）Calculate the rotation radius of the two spindles. The rotation radius of the two spindles is calculated according to the following formula:

$$i_x = \frac{l_{ox}}{\lambda} \quad (2)$$

$$i_y = \frac{l_{oy}}{\lambda} \quad (3)$$

i_x — Radius of gyration to X axis;

l_{ox} — Calculated length in x-axis direction;

λ — Slenderness ratio;

i_y — Radius of gyration to Y-axis;

l_{oy} — Calculated length in Y-axis direction.

$l_{ox} = l_{oy} = 5m$, $\lambda = 100$, it can be calculated that $i_x = i_y = 0.05m = 5cm$.

3）Cross sectional area A and radius of gyration i_x,$i_y$are obtained in the previous two steps.It is impossible to select from the list of common H-shaped steel to meet the requirements of A_{req}, i_{xreq}, i_{yreq} at the same time.We can choose mainly A_{req}, i_{yreq} and supplemented byi_{xreq}.According to the table, the size should be $250 \times 250 \times 9 \times 14$ (mm) HW section steel with the same size as the model.

4）Section checking calculation: the strength, stiffness and stability of steel frame column of this size has been checked in the previous chapter and meets the specification requirements, so the $250 \times 250 \times 9 \times 14$ (mm) HW section steel is the optimal size of the column.

Although the column section stress is surplus, in order to meet the requirements of overall stability, we can not replace it with a smaller section.

5.2. Optimization of top purlin

The top purlin and beam are mainly subjects to bending tension, without bending instability or bending torsional buckling. The axial tension and the section maximum stress is small. Therefore, the round pipe bars can be selected for the top purlins, with the size of $\varphi 114 \times 4$ (mm) , the section parameters are shown in the figure3.

By modifying the section type data in ANSYS software, it can be calculated that the maximum tensile stress of the section of the top purlin and beam is 284050pa, and it appears in the cross beam at the connection of side plate and curved plate, which is less than the tensile strength design value of Q235 steel. The cross-sectional area of round tubular steel is 13.82cm^2,while it is 92.18cm^2 for the H-shaped beam.The area is reduced by 78.36cm^2, a decrease of about 85%.After optimization, not only steel is saved ,but also self weight is reduced.

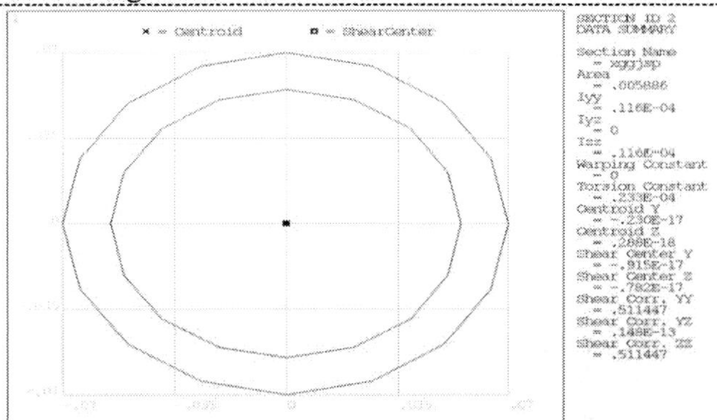

Figure 3. Cross section of top purlin round tubular steel

6. Conclusion

1) According to the load code for buildings and structures, the displacement and stress distribution of the structure under five working conditions: self weight, snow load (including uneven distribution), wind load, wind snow combination and seismic load are calculated. The calculation results under five working conditions are summarized and analyzed to determine the most unfavorable working condition and the weak positions. The most unfavorable working condition is the combination of wind and snow load, and the weak positions include the bottom of the columns and the steel frame at the curved panel.

2) The strength, stiffness and local stability of the steel frame under the most unfavorable working condition and the bearing capacity limit state are checked, which meet the requirements of the code.

3) Combined with the calculation and analysis results and ANSYS software, the barrier steel frame structure is designed and optimized. In order to meet the requirements of overall stability, the column cannot be replaced with a smaller section. The top purlin and cross beam can replace the H-section steel with the round tubular one, so as to achieve the purpose of material saving and economy.

References

[1] Huang J, Guo B, Guo X 2015 Research progress on the impact of traffic noise on population health *Journal of Peking University (Medical Edition)* **47** (03) 555-8
[2] Song S 2014 Noise reduction effect analysis and model selection applicability of Expressway sound barrier *Chang'an University*
[3] Zhu F 2019 Study on noise reduction method of Expressway under expansion *Southeast University*
[4] Liu X 2010 Dynamic characteristic analysis of large sound barrier structure *Dalian Jiaotong University*
[5] Wang T, Feng M, Xu J 2012 Design and construction of fully enclosed sound barrier for urban viaduct *Environmental engineering* **30**(03) 113-6
[6] Ma C 2017 Study on static and fatigue performance of large fully enclosed sound barrier for freight railway *Huazhong University of science and technology*

Study on Structural Reliability of Prestressed Reinforced Concrete Beam Bridges Based on Full Probability Analysis

Yubao Liu[1,*], **Hanxiong Liu**[2] and **Shouyi Sun**[3]

[1]CCCC First Harbor Engineering Co. Ltd. Tianjin, China
[2]CCCC Second Highway Consultants Co., Ltd. Wuhan, Hubei, China
[3]School of transportation and logistics engineering, Wuhan University of technology, Wuhan, Hubei, China

Email: 2321547575@qq.com

Abstract. A three dimensional(3D) finite element model is established for a prestressed concrete girder bridge, list the limit state equation and variable probability distribution model, obtain the failure probability and reliability index of the limit state equation based on the Monte-Carlo method, investigate the reliability index's sensitivity to random variables for limit state equations of bridge bearing capacity. The specific method of the research on the structure reliability index calculation is to select a suitable way that can fit the research's object among the center point method, the checking-point method, the Monte-Carlo method, the importance sampling method, and so on. Authors use finite element software ANSYS to build model and perform 3D force analysis. According to the results, authors can determine the bridges' failure modes under the state of 3D stress. Then, authors can list the possible limit state equations related to Stiffness, strength, function, durability, and some character else. After determining the bridge's structural limit state equation and the probability distribution model of the variables, determine the distribution function of each impact factor and its characteristic parameters through the site survey data and other related data surveys. Then, use the Monte-Carlo method for the calculation of bridge reliability index, obtain the failure probability and reliability index of each limit state equation, analyze the sensitivity of each variable to the reliability index under the ultimate state of the bridge's bearing capacity. At last, authors give construction quality control plans and suggestions according to the data above.
Keywords. Bridge engineering; Full probability analysis; Structural reliability; Prestressed Bridges; Reinforced concrete Bridges

1. Introduction

According to the statistics, as of the end of 2006, there were 6,282 dangerous highway bridges in the fifth category of health in our country. Among of them , After the pre-stressed concrete bridge is built and put into use for a period of time, the bridges' structural materials will have the problem of natural aging, coupled with the increasing vehicle load and improper use and maintenance, the bridges will have various structural damages and defects. These will bring difficulties to the corresponding maintenance and management of the bridge. Therefore, how to evaluate the reliability of bridge operations and carry out corresponding maintenance and management have become a topic of concern to the domestic and foreign bridge workers

The reliability analysis example of PC beam bridge shows: evaluate the safety status of the reinforced concrete bridges in service scientifically, determine the reliability of bridges from different aspects and provide a practical basis for highway and bridge operation management unit will help

IOP Conf. Series: Materials Science and Engineering 1210 (2021) 012003 doi:10.1088/1757-899X/1210/1/012003

Construction Department take reasonable reinforcement methods for the bridge to restore and improve the bridge's bearing capacity.

Until 2012, 43.88 billion yuan was invested in the renovation of dangerous bridges across the country and 21,610 dangerous bridges with a total of 1.741 million meters were retrofit. To replace these seriously diseased old bridges will cost a lot of money and the traffic will be closed. This situation is not allowed by our national conditions and financial resources. Although the development of the reliability theory of prestressed concrete beam bridge structure has a considerable scale at present, there is still quite a large gap in the application of reliability analysis theory to the engineering practice. Therefore, the relevant research is needed to continue.

At present, in the relevant specifications, for prestressed concrete structural members designed by limit state control. The target and reliable indicators of its crack resistance (crack width), deflection and stress are not clear yet. Although the limits of the above indicators are given in the design specifications, their reliability levels have not been determined. Therefore, it's our duty to establish a reliability analysis model of prestressed concrete bridge structure considered the ultimate state of bearing capacity and the limit state of normal use. Apart from this, understand the reliability level of various indicators of prestressed concrete bridge members designed in accordance with the country's standards and analyze the influencing parameters are also necessary.

The research objects of this paper include the basic principles and calculation methods of the reliability of prestressed reinforced concrete structures, as well as the probability distribution types of the commonly used variables. By using the finite element software ANSYS to establish a model, authors completely analyze Reinforced concrete beam bridge and give the construction quality control plan and suggestions to the relevant departments

2. Structural reliability

The probability of the structure or the structural member complete a predetermined function is called the reliability probability, also can be called reliability P_s. The probability that the structure cannot complete its intended function is called the failure probability. It also can be expressed as P_f. And assuming that the basic random variable in the structure is $x_1, x_2, \ldots, x_n$, the corresponding probability density function is $f_x(x_1, x_2, \ldots, x_n)$, the functional function represented by these random variables is

$Z = g_x(x_1, x_2, \ldots, x_n)$. Then the failure probability of the structure can be expressed as

$$P_f = P(Z < 0) = \iint_{z<0} \cdots \int f_x(x_1, x_2, \ldots, x_n) dx_1 dx_2 \ldots dx_n \tag{1}$$

Assuming that the random variable of resistance of the structure is R and the random variable of the load is S. The corresponding probability density functions and probability distribution functions are $f_R(r)$ and $f_S(s)$.The probability distribution functions are $F_R(r)$ and $F_S(s)$. R and S are independent of each other. Then the probability of structural failure is:

$$P_f = \iint_{R<S} f(r, s) dr ds \tag{2}$$

If R and S are normal distribution. their mean and standard deviation are μ_R, μ_S and σ_R , σ_S .Then Z is also a normal random variable. Then its mean is $\mu_z = \mu_R - \mu_s$ and its standard deviation is

$$\sigma_Z = \sqrt{\sigma_R^2 + \sigma_S^2} \tag{3}$$

And $\beta = \frac{\mu_Z}{\sigma_Z}$ is the reliable index

3. Failure Mode and Limit State Equation

The limit state is a sign which can distinguish the success and the failure of the structural working state. In the course of use, if the entire structure or part of the structure exceeds a specific state and cannot meet a certain functional requirement specified by the design, this specific state is called the limit state of the function. And the limit state can be divided into two types: the bearing capacity limit state and the normal service limit state.

3.1. Load capacity limit state.
The ultimate state of bearing capacity refers to the structure or the structural member reaching the maximum bearing capacity, and then there will be fatigue, overturning, instability, drift, continuous collapse and other failures or deformation that is not suitable for continued bearing. The design expression for the ultimate state of the bearing capacity is as follows:

$$\gamma_0 S \leq \mathrm{R}$$

$$\mathrm{R} = \mathrm{R}(f_c, \ f_s, \ \alpha_K \ldots) \tag{4}$$

In the formula, γ_0 is the important factor of the structure. S is the design value of the effect combination under the limit state of the bearing capacity. R is the design value of the resistance of the structural member. γ_{Rd} is the uncertainty coefficient of the structural member resistance model. f_c, f_s is the design value of concrete and steel bar strength. α_k is the standard value of the geometric parameter.

3.1.1. Variable load effect control combination.

$$\mathrm{S} = \gamma_G S_{GK} + \gamma_{Q1} S_{Q1K} + \sum_{i=2}^{n} \Phi_{ci} \gamma_{Qi} S_{Qik} \tag{5}$$

3.1.2. Combination of permanent load effect control.

$$\mathrm{S} = \gamma_G S_{GK} + \sum_{i=1}^{n} \Phi_{ci} \gamma_{Qi} S_{Qik} \tag{6}$$

In the formula, γ_G is the partial coefficient of permanent load, with a numerical value of 1.2. γ_{Q1} and γ_{Qi} is the partial coefficient of the first and i-th variable load, with a numerical value of 1.4. S_{GK} is the load effect value calculated according to the permanent load standard value G_K. S_{Qik} is the load effect value calculated according to the variable load standard value Q_{ik} . φ_{ci} is the wind load with a value of 0.6, other coefficients are with the value of 0.7. n is the variable number of loads participating in the combination.

3.2. Normal use limit state.
The normal use limit state refers to a certain limit corresponding to the normal service and durability performance of the structure or structural member. The design expression for the serviceable limit state is: $S_d < C$

In the formula, S is the design value of load effects such as deformation, cracks, and so on. C is the limit for structural members to reach the deformation, stress, crack width and natural frequency specified in the normal use requirements.

Taking S_d under the standard combination, the structure function function is:

$$\begin{aligned} Z &= R - S \\ &= C - S_d \\ &= \mathrm{C} - (S_{GK} + S_{Q1K} + \textstyle\sum_{i=2}^{n} \Phi_{ci} S_{Qik}) \end{aligned} \tag{7}$$

Taking S_d under the permanent combination, the structure-function is:

$$\begin{aligned} Z &= R - S \\ &= C - S_d \\ &= \mathrm{C} - (S_{GK} + S_{Q1K} + \textstyle\sum_{i=2}^{n} \Phi_{ci} S_{Qik}) \end{aligned} \tag{8}$$

4. Reliability analysis and calculation

4.1. The flexural limit state equation

$$\mathrm{Z} = \mathrm{R} - \mathrm{S}$$

$$= f_{cd}(b_f' - b)h_f'\left(h_0 - \frac{h_f'}{2}\right) + f_{cd}bx\left(h_0 - \frac{x}{2}\right) - \gamma_0 S_{ud}$$

$$= f_{cd}(b_f' - b)h_f'\left(h_0 - \frac{h_f'}{2}\right) + [f_pA_p - f_{cd}(b_f' - b)h_f']h_0 - \frac{[f_pA_p - f_{cd}(b_f' - b)h_f']^2}{2f_{cd}b} - \gamma_0 (M_{DI} + M_{DII} + M_L + M_t) \tag{9}$$

4.2. The shear limit state equation

$$Z = R - S$$
$$= V_u - \gamma_0 V_d$$
$$= \alpha_1\alpha_2\alpha_3(0.45 \times 10^{-3})bh_0\sqrt{(2 + 0.6p)\sqrt{f_{cu,k}\rho_{sv}f_{sv}}} + (0.75 \times 10^{-3})f_{sd}\sum A_{sb}sin\theta_s - \gamma$$
$$= \alpha_1\alpha_2\alpha_3(0.45 \times 10^{-3})bh_0\sqrt{(2 + 0.6 \times 100)\frac{(A_P + A_{pb} + A_a)}{bh_0}\sqrt{f_{cu,k}\frac{A_{sv}}{S_v b}f_{sv}}}$$
$$+(0.75 \times 10^{-3})\, f_{sd}\sum A_{sb}sin\theta_s - \gamma_0 V_d \tag{10}$$

4.3. Crack resistance limit state equation

$$Z = [W_{fk}] - W_{fk}$$

$$= [W_{fk}] - C_1C_2C_3 \times \frac{\frac{M_S + N_{P0}(Z - e_p)}{(A_P + A_S)Z}}{E_S} \times \frac{30 + d}{0.28 + 10 \times \frac{A_S + A_p}{bh_0 + (b_f' - b)h_f}}$$

$$= [W_{fk}] - C_1C_2C_3 \times \frac{\frac{M_S + [(\sigma_{con} - \sigma_l)A_P - \sigma_{16}A_S](Z - e_p)}{(A_P + A_S)Z}}{E_S} \times \frac{30 + d}{0.28 + 10 \times \frac{A_S + A_P}{bh_0 + (b_f' - b)h_f}} \tag{11}$$

4.4. Deflection limit state equation

$$Z = R - S$$
$$= \frac{l}{600} - (f_{sl} - f_{gl})$$
$$= \frac{l}{600} - \eta_\theta\,(f_s - f_g)$$
$$= \frac{l}{600} - \eta_\theta \times \frac{5l^2}{48 \times 0.95E_cI_0}\,(M_S - M_G) \tag{12}$$

4.5. Reliability analysis results.

Authors use a calculation program compiled by MATLAB to calculate the reliability of the deflection according to the Monte Carlo method, combine the existing literature and the statistical parameters in data statistics according to the limit state equation of the deflection of the prestressed concrete bridge structure defined above. Then, authors can reveal and analyze the reliability level of prestressed concrete simply supported beams designed according to the current specifications. Furthermore, authors use Matlab to check the calculation, get the corresponding reliability index and compare it with the target reliability index required by the relevant specifications at last.

The following contents are part of the code of the flexural reliability program and the target reliability index during the bridge design period which is shown in the Table1:

```
muX=[1312.16; 4587; 25.81; 2403.7739; 202.4263; 202.3524; 1822.6324];
cvX=[0.01; 0.008; 0.0344; 0.0029; 0.0004; 0.0056; 0.0021];
sigmaX = cvX * muX
nS=5e6;  ig=ones(nS:1)
n=length(muX)
for k=1; n,x(:,k)=normrnd(muX(k),sigmaX(k),nS,1); end
t=x(:,1).*x(;,2)/(x(:,3).*x(;,4));
if t<x(;,6)
g=x(;,1).*x(;,2)./(x(;,7)-x(;,1)*x(;,2)./(2.*x(;,3).*x(;,4)))...
-100043.34e6
else
g=x(;,3).*(x(;,4)-x(;,5)).*x(;,6).*(x(;,7)-x(;,6)./2)+...
x(;,7).*(x(;,1).*x(;,2)-x(;,3).*x(;,4)-x(;,5)).*x(;,6))...
-(x(;,1)*x(;,2)-x(;,3).*(x(;,4)-...
x(;,5)).*x(;,6)).^2./(2.*x(;,3).*x(;,5))-10043.34e6;end
nF=sum(ig(g<0));
```

Table 1. Target reliability index during the bridge design period

Combination of action and effect	1st level		2nd level		3rd level	
	ductile damage	Brittle damage	ductile damage	Brittle damage	ductile damage	Brittle damage
Main combination	4.7	5.2	4.2	4.7	3.7	4.2
extra combination	4.2	3.7	3.7	4.2	3.2	3.7

5. Random Variable Sensitivity Analysis

Sensitivity analysis is while a random variable changes within a certain range, authors can study the impact of changes in these variables on the model results. And the main purpose of sensitivity analysis is to obtain the sensitivity coefficient of each variable by analyzing the relevant variables of the model. As examples, the sensitivity of the reliability index to the average value of f_p and the sensitivity of the reliability index to the coefficient of variation of f_p are shown in the Figure1 and the Figure2.

In the actual work, researchers can ignore random variables with small sensitivity coefficients based on our work experience. By treating some random variables as constants, the complexity of the model can be greatly reduced and the workload of data analysis can be reduced. We need to focus on random variables with large sensitivity coefficients, and strictly control them during the design, so as to ensure that the structure is safe enough. Under such measures, the work efficiency is improved without affecting the calculation accuracy.

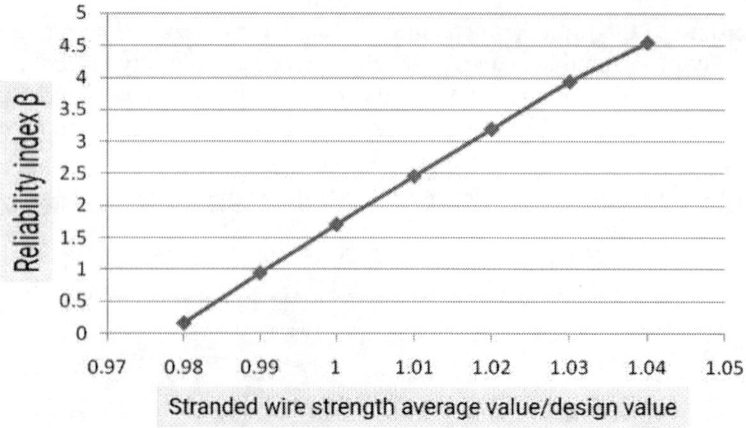

Figure 1. The sensitivity of the reliability index to the average value of f_p

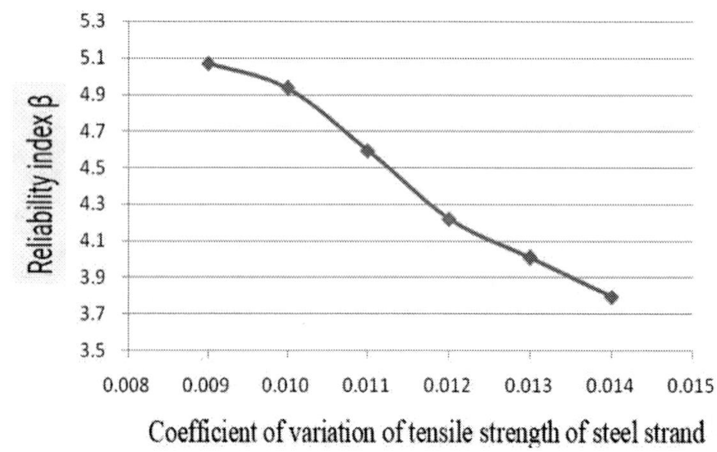

Figure 2. The sensitivity of the reliability index to the coefficient of variation of f_p

By changing the mean value and the coefficient of variation δ of the resistance reliability random variables, sensitivity analysis is performed on the seven random variables in the expression. Finally, a summary of the results can be got as shown in the Table2 and the Table3:

Table 2. Random Variable Mean Sensitivity

	f_p	A_P	f_{cd}	b'_f	b	h'_f	h_0
Average percentage increase(%)	6	6	37.5	37.5	-	-	6
Reliability index β increase percentage(%)	2800	6207	21	18	-	-	1991

Table 3. Random Variable Coefficient of Variation Sensitivity

	f_p	A_P	f_{cd}	b'_f	b	h'_f	h_0
Percentage increase in the coefficient of variation(%)	56	57	60	55	-	-	57
Reliability index β reduce percentage(%)	34	22	4	4	-	-	10

According to the table above, some constructive suggestions can be given as follows:

- 1)Strictly control the performance of prestressed steel bar materials (tensile strength, area) during design and construction
- 2)PC beam web width b and roof thickness h_f' can be treated as constants in resistance analysis, and the requirements can be appropriately relaxed during construction acceptance.

6. Conclusion

This subject is adopted for the first time that the total probability method is used to analyze the reliability of the prestressed reinforced concrete beam bridge structure with multiple failure modes. Authors combine the existing structural finite element analysis software and Monte Carlo numerical simulation program to solve the establishment of the relevant model of the random process of each variable and the solution of the equation.

In actual engineering applications, on the basis of reliability basic theory and method research foundation, authors study the failure modes and limit states of the prestressed concrete beam bridges and analyze the sensitivity of each random variable to the reliability index under the ultimate state of the bridge's bearing capacity. Finally, authors can use these as the basis to obtain cost-effective construction quality control plans and suggestions which meet the requirements.

References

[1] Du B 2009 Research on reliability and remaining service life of existing presressed concrete bridge *Chengdu: Southwest Jiaotong University*

[2] Zhao G 1996 *Theory and application of engineering structure reliability* (Dalian: Dalian University of Technology Press)

[3] Gong J, Zhao G 2005 Application and development of structure reliability theory aboard *China Civil Engineering Journal* **38**(2) 1-7

[4] Zhao G, Gong J, Zhao S 2000 Developments of reliability research for civil engineering structure in China *China Civil Engineering Journal* **40**(3) 253-8

[5] Yao J, Zhao G, Pu Y 2000 Independent Increment Process Model for Stochastic Structual Resistance ——*Proc. of the Ninth Ann. Conf. of China Civil Engineering Society* (Beijing: China Water Conservancy and Hydropower Press)

[6] Lu Deng, C.S. Cai 2011 Reliability-based dynamic load allowance for capacity rating of presressed concrete girder bridges *Bridge Eng* **16** 872-80

[7] Li Y, Bao W, Guo X, Cheng X 1997 *Reliability and probability limit state design of highway bridge structure* (Beijing: China Communications Press)

[8] Dimitri V V, Leonid C 2009 Serviceability Reliability of Reinforced Concrete Beams with Corroded Reinforcement *Struct.Eng* **135** 896-905

An Intelligent Bearing Fault Diagnosis Method Based on SF-SVM

Bao'an Qiu, Pan Sun* and Lili Li

The 14th Research Institute of China Electronics Technology, Nanjing, Jiangsu, China

*corresponding author's e-mail: 562251022@qq.com

Abstract. Rolling bearing, as a key component of rotating machinery, its health status directly determines the stability and reliability of the whole machine. The research on its intelligent diagnosis method has important engineering value and academic significance. However, due to actual engineering conditions, the types of bearing failures and the amount of data are limited. Aiming at the difficulty of extracting and selecting bearing vibration features under limited sample constraints, this pa-per proposes an intelligent fault diagnosis method of SF-SVM. On the basis of the short-time Fourier change, the L2 regularized sparse filter is used to extract the unsupervised feature of the bearing vibration time-frequency map. After obtaining the typical features of the bearing, the support vector machine is used for diagnosis.

1. Introduction

With the rapid development of perception technology and information technology, people's de-mand for digitization, informatization and intelligence of rotating machinery is higher and high-er. Rolling bearing, as a key component of rotating equipment. Its' health state during operation directly determines the stability and reliability of equipment[1,2]. According to statistics, the pro-portion of rotating machinery faults caused by rolling bearing failure is 45% ~ 55%[3].

The research on the fault diagnosis method of rolling bearing can provide guidance for the pre-dictive maintenance of bearing and overall equipment, prolong the service life of bearing and reduce the operation and maintenance cost of bearing, which has important engineering value and academic significance.

During the operation of the bearing, the signals obtained by the sensor in real time are high-dimensional and large quantities, but limited by the actual situation, the type and number of faults are limited. Traditional diagnosis methods rely too much on expert experience for feature extraction and feature selection, and deep learning methods need a lot of data for training and learning. Therefore, aiming at the difficulty of bearing vibration feature extraction and selection under small samples, this paper proposes a fault diagnosis method of SF-SVM.

2. Research status at home and abroad

Bearing fault diagnosis is to extract features from the collected signals to judge the health status of bearings. Traditional fault diagnosis methods include KNN [4-6], support vector machine SVM [7,8], neural network [9,10], etc. W. Qin et al. [7] used IMF information entropy as the feature input of SVM to realize the fault diagnosis of rolling bearing. Tang G [11] solved the problem of effectively extracting fault features and selecting high-precision classifier, and pro-posed a rolling bearing fault diagnosis method based on improved fast spectral correlation and optimized random forest.

With its strong expression ability and hierarchical learning ability, deep learning algorithm has developed rapidly in the fields of speech recognition and target detection, and is also favored by researchers in the field of rotating machinery fault diagnosis. Aiming at the problem that tradi-tional feature extraction depends on expert experience and prior knowledge, Jia f et al.[12] pro-posed a fault diagnosis method based on deep neural network (DNN). Olivier Janssens et al.[13] proposed a feature learning model based on convolutional neural network, which directly learns the typical features used for bearing state judgment from the original data. Qi y et al.[14] proposed a fault diagnosis method based on stacked sparse self encoder, which can extract more repre-sentative high-dimensional features from bearing vibration signals. Gan m et al.[15] extracted the fault characteristics of bearing signals with the help of wavelet packet transformation and used them as the input of deep confidence network to realize the fault diagnosis of rolling bearing.

To sum up, traditional methods need to use a priori knowledge and engineering experience for feature extraction, while deep learning methods can be fully trained on the premise of enough training samples. Therefore, it is necessary to study the bearing diagnosis method under the con-dition of limited samples.

3. Fault diagnosis method based on SF-SVM

The algorithm framework of fault diagnosis is shown in Fig.1, including signal preprocessing, feature extraction and fault diagnosis. Signal preprocessing is the preliminary processing of the collected signals, including signal truncation, short-time Fourier transform, division of training set and test set, etc. Feature extraction, that is, using the sparse filtering method to extract the features with sparsity within and between samples from the time-frequency map, so as to reduce the dimension of the signal. In fault identification, under the constraint of limited samples, soft interval support vector machine is selected to suppress over fitting by introducing relaxation variables into the loss function.

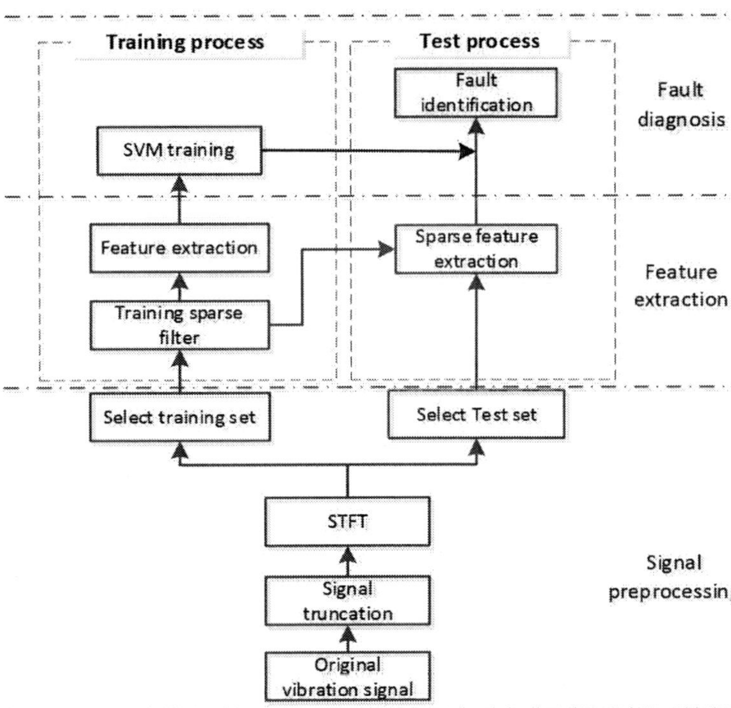

Figure 1. Algorithm framework

3. 1. Signal preprocessing - short time Fourier transform

The standard Fourier transform has the problem that the frequency domain information and time domain information cannot be localized at the same time. The short-time Fourier transform di-vides the signal into many small time intervals. Analyzing each time interval with Fourier trans-form can master the local frequency of the signal and the time period information locked out of the frequency.

Given a signal $x(t) \in L^2(R)$, its STFT is defined as follows.

$$STFT_x(t,\Omega) = \int x(\tau) g(\tau - t) e^{-j\Omega\tau} d\tau \quad (1)$$

The meaning of this formula is to intercept the signal x (T) with the window function g (T) in the time domain, and Fourier transform the intercepted local signal. By continuously moving the center of the window function, the Fourier transform at different times can be obtained. The set of these Fourier transforms is $STFT_x(t,\Omega)$.

3. 2. Feature extraction

3. 2. 1. Sparse filtering principle.Sparse filtering principle. Sparse filtering is an unsupervised feature learning method proposed by Ngiam[16], which extracts the features of high dispersion and population sparsity by optimizing the L1 / L2 norm of the feature matrix. Suppose the sample data is $X=[x_1,x_2,..,x_i,...,x_M]$, where $x_i \in R^N$, i.e. x $\in R^{NxM}$, and its corresponding weight matrix is w $\in R^{N_f xN}$, then the feature matrix F corresponding to F = Wx, where F $\in R^{N_f xM}$. The feature matrix is expressed as F = $[f_1,f_2,..,f_i,...f_M]$, where f_i represents feature vector corresponding to sample data x_i and it can be calculated as follows.

$$f_i^{(j)} = w^{(j)} \times x_i \quad (2)$$

Then, the same features of all training samples are normalized by L2 norm, and the feature vector after L2 norm normalization of the j-th feature of all samples $\dot{f}^{(j)}$ is recorded as follows.

$$\dot{f}^{(j)} = \frac{f^{(j)}}{\| f^{(j)} \|_2} \quad (3)$$

The feature matrix $\dot{F}$ is obtained after row normalization, and then its each column is normalized by L2 norm, that is, all transformed features of the ith sample are normalized by column, and the transformed sample vector is recorded as $\hat{f}_i$.

$$\hat{f}_i = \frac{\tilde{f}_i}{\| \tilde{f}_i \|_2} \quad (4)$$

Finally, the characteristic evaluation function, namely the cost function, is obtained.

$$\min_{\arg W} \sum_{i=1}^{M} \| \hat{f}_i \|_1 = \sum_{i=1}^{M} \| \frac{\tilde{f}_i}{\| \tilde{f}_i \|_2} \|_1 \quad (5)$$

The loss function can be solved by finite memory quasi Newton method (L-BFGS).

3. 2. 2. Feature extraction based on sparse filtering. Sparse filtering is used for unsupervised feature extraction to obtain the typical features of bearings with dispersion and sparsity. In order to improve the feature extraction ability of sparse filter under sample limited constraints, L2 regularization is carried out, that is, L2 norm of feature mapping matrix is introduced into its objective function W, as shown below.

$$\min_{\arg W} \sum_{i=1}^{M} \| \frac{\tilde{f}_i}{\| \tilde{f}_i \|_2} \|_1 + \lambda \| W \|_2 \quad (6)$$

3. 3. Fault diagnosis

In order to alleviate the over fitting phenomenon of support vector machine under limited samples and noise interference, soft interval support vector machine is select-ed to introduce relaxation variables for each sample to characterize the degree that the sample does not meet the constraints. The corresponding objective function is as follows.

$$\min_{w,b} \frac{1}{2} \| w \|^2 + C\sum_{i=1}^{m} \xi_i \tag{7}$$

C > 0 is called the penalty parameter. While maximizing the interval, the number of samples that do not meet the constraints is required to be as small as possible.

4. Experimental verification

4. 1. Experiment introduction

This experiment selects the public data set of the bearing data center of Case Western Reserve University [17]. As shown in the figure 2, the bearing fault simulation test-bed is composed of drive motor, frequency converter, load motor, vibration ac-celeration sensor, collector, etc. There are 4 test conditions in total, as shown in Table 1 below.

In this experiment, the fault of the bearing is simulated by machining defects of dif-ferent scales at different parts of the bearing. The specific types are shown in Table 1. The fault of the bearing is expressed in the form of xx-x, where XX represents the fault location, if represents the inner ring fault, of represents the outer ring fault, Ba represents the rolling element fault, X represents the size of the fault scale, and the larger the number, the larger the fault scale.

Figure 2. Experiment table

Table 1. Bearing fault parameters.

Fault type	Fault scale (mm)	Fault symbol representation
inner race/ outter race /ball	1.78	IF-1/OF-1/BA-1
inner race/ outter race /ball	3.56	IF-2/OF-1/BA-2
inner race/ outter race /ball	5.15	IF-3/OF-3/BA-3

4. 2. Parameter selection

The method proposed in this paper needs to select several parameters, including the length of input samples, the number of extracted features, sparse filtering and regularization parameters λ, Penalty coefficient C of support vector machine. Therefore, the parameter selection was studied, and each test was repeated 30 times to reduce the influence of random factors.

4. 2. 1. Sample length. In order to study the influence of sample length, that is, the input dimension of sparse filtering, on the diagnosis results, 0.5% samples are randomly selected for model training and regulariza-tion parameters λ= 0.01, penalty coefficient C = 0.1, the number of features extracted by sparse filtering is set to 100, and the diagnosis results are shown in Figure 4. When the sample length is within [10005000], the diagnostic accuracy and stability are almost unchanged, indicating that the sample length has little impact on the diagnostic results. In this paper, the sample length is set to 1000.

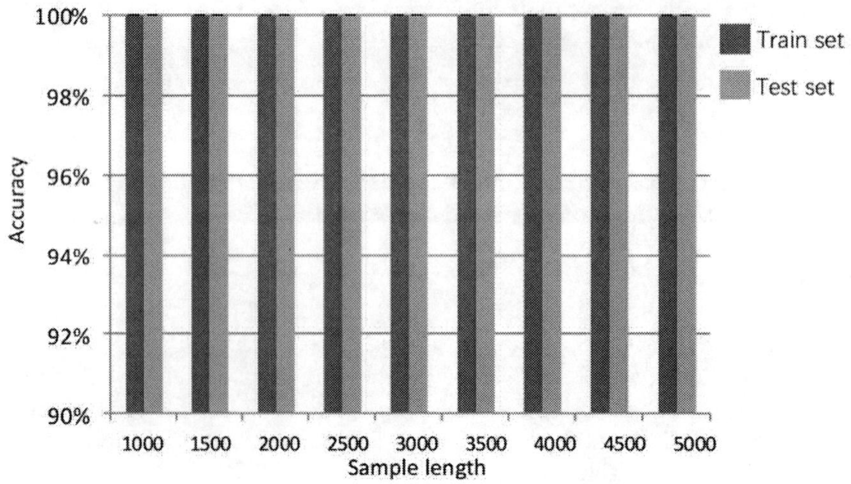

Figure 3. The influence of sample length

4. 2. 2. Feature number. The influence of the number of features extracted by sparse filtering on the diagnosis results is shown in Fig.4 With the gradual increase of the number of features extracted, the diagnosis ac-curacy and stability are also gradually improved. When the number of features is greater than 30, the diagnosis tends to be stable, so this paper is set λ= 30.

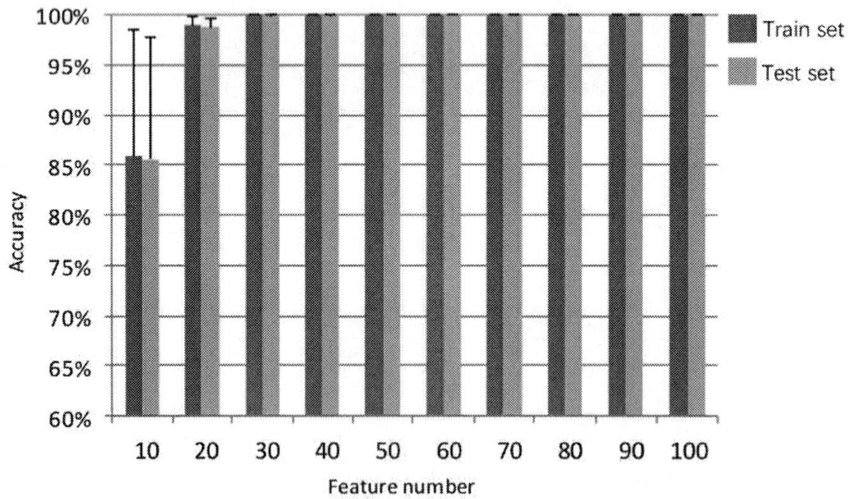

Figure 4. The influence of feature number

4. 2. 3. Parameter λ. Sparse filter regularization parameters λ The influence on the diagnosis results is shown in Fig.5, with the regularization parameters λ The diagnostic accuracy and stability are also gradually im-proved λ When ≥ 0.01, the diagnosis tends to be stable, so it is set in this paper λ= 0.01.

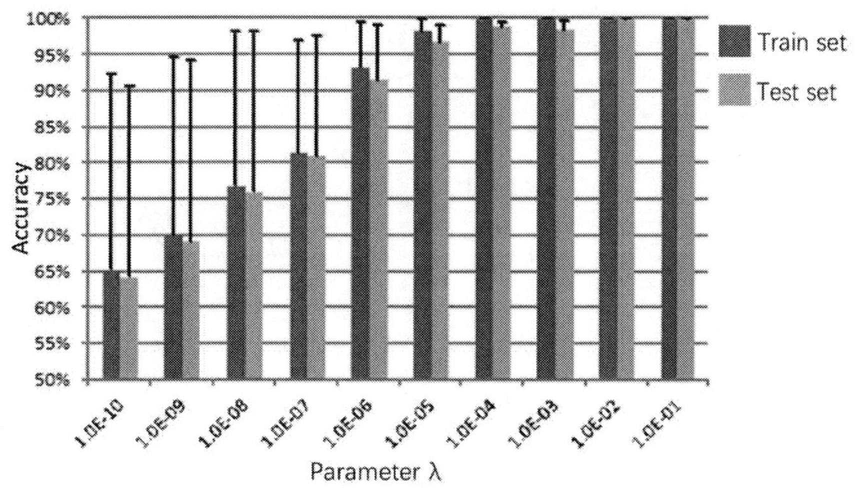

Figure 5. The influence of parameter λ

4. 2. 4. Parameter C. The influence of the penalty coefficient C on the diagnosis result is shown in Figure 6. When C is in the [0,0.1] range, the diagnosis accuracy and stability are almost unchanged with the change of C, indicating that the penalty coefficient C has little influence on the diagnosis result, so C = 0.1 is set in this paper.

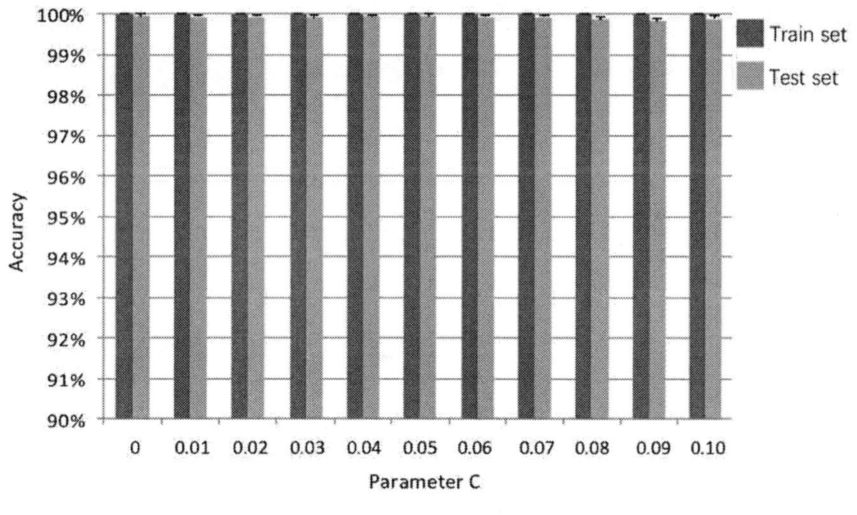

Figure 6. The influence of parameter C

4. 2. 5. Train set ratio. The influence of the proportion of training set on the diagnosis results is shown in Fig.7. When the proportion of training set is in the interval of [0.01,0.1], the diagnosis accuracy and stability gradually improve with the increase of the proportion of training set. When the proportion of training set is greater than 0.05, the diagnosis accuracy tends to be stable, indicating that the model is trained more fully with the increase of training sample data.

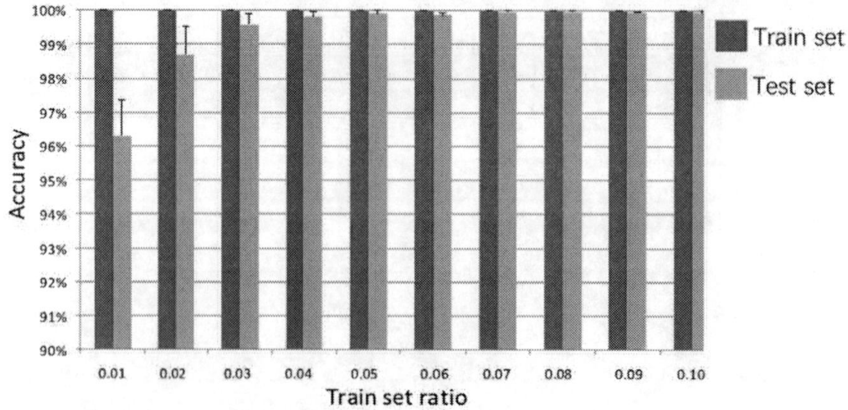

Figure 7. The influence of train set ratio

4. 3. Comparison of results

In order to verify the fault diagnosis model in this paper, the diagnosis methods proposed in this paper are compared with the existing methods, as shown in Table 2, from the ratio of training samples, operating conditions of bearings, number of bearing States, recognition rate, etc.

Compared with the results of the existing literature, this method has better recognition perfor-mance under complex working conditions with less training samples.

Table 2. Comparison of diagnosis results

Method	Condition	Train set ratio	Number of bearing types	Accuracy
Wavelet+SVMS[18]	1	75%	10	88.9%
TR-LD[19]	4	10%	10	92.6%
SF+softmax[20]	1,2,3,4	10%	10	99.7%
Proposed Method	1,2,3,4	**0.1%**	10	**99.9%**

5. Conclusion

Aiming at the difficulty of bearing vibration feature extraction and fault recognition under lim-ited sample constraints, this paper proposes a fault diagnosis method of SF-SVM, uses L2 regu-larized sparse filter to automatically extract the representative features of bearing from time-frequency diagram, and realizes fault diagnosis by support vector machine. Compared with the existing literature, the superiority of this method is verified.

References

[1] Qi X X, Ji J W, Han X W 2011 Fault Diagnosis Methods of Rolling Bearing: A General Re-view *Key Engineering Materials* 480-1+986-92

[2] S. Tyagi, et al 2017 A DWT and SVM based method for rolling element bearing fault diagnosis and its comparison with Artificial Neural Networks *Journal of Applied and Computation-al Mechanics* **3**(1)

[3] Rai A, Upadhyay S H 2016 A Review on Signal Processing Techniques Utilized in the Fault Di-agnosis of Rolling Element Bearings Tribology International **96** 289–306

[4] Cai E., et al 2015 Fault diagnosis method based on LMD and KNN algorithms for rolling bear-ing *Modern Electronics Technique*

[5] Lu D, Qian N and Yang X 2018 Fault Diagnosis of Rolling Bearing Based on KNN-Naive Bayesian Algorithm *Computer Measurement & Control*

[6] Khelil J, Ramdani M, Khelil K, et al 2019 Bearing Faults Diagnosis Using Discrete Wavelets and Artificial Intelligence Approaches//*2019 1st International Conference on*

Sustain-able Renewable Energy Systems and Applications (ICSRESA)

[7] Qin W, Zhang W and Lu C 2016 Rolling bearing fault diagnosis: A data-based method using EEMD, information entropy and one-versus-one SVM//*2016 12th World Congress on In-telligent Control and Automation (WCICA), Guilin* 1016-20

[8] Huo Z, Zhang Y, Shu L and Gallimore M 2019 A New Bearing Fault Diagnosis Method Based on Fine-to-Coarse Multiscale Permutation Entropy, Laplacian Score and SVM. *IEEE Access* **7** 17050-66

[9] Gunerkar R S, Jalan A K & Belgamwar S U 2019 Fault diagnosis of rolling element bearing based on artificial neural network *Journal of Mechanical Science and Technology* **33**(2) 505–11

[10]Al-Raheem K F, et al 2008 Application of the Laplace-Wavelet Combined With ANN for Rolling Bearing Fault Diagnosis *International journal of comadem* **11**(4) 18-24.

[11]Tang G; Pang B; Tian T, Zhou C 2018 Fault Diagnosis of Rolling Bearings Based on Im-proved Fast Spectral Correlation and Optimized Random Forest *Applied Science* **8**(10) 1859

[12]Jia F, Lei Y, Lin J, et al 2016 Deep neural networks: A promising tool for fault characteristic mining and intelligent diagnosis of rotating machinery with massive data *Mechanical Systems and Signal Processing* **72-73** 303-15

[13]Olivier Janssens, Viktor Slavkovikj, Bram Vervisch, Kurt Stockman, Mia Loccufier. 2016 Con-volutional Neural Network Based Fault Detection for Rotating Machinery *Journal of Sound and Vibration* **377** 331-45

[14]Qi Y, Shen C, Liu J, et al 2017 An Automatic Feature Learning and Fault Diagnosis Method Based on Stacked Sparse Autoencoder//International Workshop of Advanced Manu-facturing & Automation. Springer, Singapore **7** 367-75

[15]Gan M, Wang C, Zhu C 2016 Construction of hierarchical diagnosis network based on deep learning and its application in the fault pattern recognition of rolling element bearings *Mechanical Systems and Signal Processing* **72**(73) 92-104

[16]Ngiam J, Pang W K, Chen Z, et al 2011 Sparse filtering *International Conference on Neural Information Processing Systems*

[17] http://csegroups.case.edu/bearingdatacenter/pages/welcome-case-western-reserve-university-bearing-data-center-website

[18]Du W, Tao J, Li Y, et al 2014 Wavelet leaders multifractal features based fault diagnosis of rotating mechanism*Mechanical Systems and Signal Processing* **43**(s 1–2) 57–75

[19]Jin X, Zhao M, Chow T W S, et al 2014 Motor Bearing Fault Diagnosis Using Trace Ratio Lin-ear Discriminant Analysis *IEEE Transactions on Industrial Electronics* **61**(5) 2441-51

[20]Lei Y, Jia F, Lin J, et al 2016 An intelligent fault diagnosis method using unsupervised feature learning towards mechanical big data *IEEE Transactions on Industrial Electronics* 1-1

AUTHOR INDEX